TAKING THE FRUSTRATION OUT OF MATH

By

Mary Hood, Ph.D.

ISBN: 0-9707597-6-2

Third Edition, 2002

Published by Ambleside Educational Press
P. O. Box 2524
Cartersville, GA 30120

Frustration

One of my clearest memories from high school days remains the morning when Mrs. Schneider, my advanced math teacher, announced a pop quiz. The class had been studying logarithms for six weeks, and although I had attended every day, I still didn't have the foggiest idea what 10 log 9 meant. I can recall the hopeless feeling that overwhelmed me when I looked over the sheet of paper. There wasn't one recognizable question on the quiz. My mind raced. Should I cover the page with meaningless numbers? Should I leave it blank? Should I come up with some clever comments and cartoon characters, and hope she was in a jovial mood?

I don't remember what I decided, but I do recall the day several weeks later when she called me into her office. She explained that I was making an "F" in her course, but that if I promised never to take a math class from her again, she'd give me a passing grade. Naturally, I started discussing how much I enjoyed math, and feigned interest in taking calculus the following fall. The horrified look on her face provided all the satisfaction I needed. That's how I got attention back in those days. If I couldn't get good grades, at least I could succeed at being the biggest smart-aleck in the class.

That wasn't the first time I had an encounter with a disgusted adult over my lack of mathematical ability. The teachers at my elementary school often berated my efforts to make sense of the black squiggles on the page in front of me. They claimed I was just being lazy and stupid. I knew I was really trying, but my brain simply turned to mush when somebody began a math problem with, "One car started out from Baltimore at 9:50 in the morning, traveling 50 m.p.h., and the other...."

As a child, I was a perfect example of someone who was totally frustrated by math. Despite testing into a gifted program in the fifth grade, and being tracked "college prep" in high school, I could never understand the simplest mathematical concepts. It took years for me to overcome this difficulty, but eventually I did. In my early twenties, I decided it was finally time to do something about it. Once the motivation was truly my own and I had achieved the proper level of readiness (which obviously came pretty late in my case!), I was able to master the basics with little difficulty.

One day when I was in my thirties, I started reminiscing about Mrs. Schneider's class and thought, "I wonder what logarithms really are?" I went to the shelf, took out an encyclopedia, and learned all about them in less than ten minutes. Of course, I also immediately forgot them, because they weren't something that I really needed to know at the time. The information went in one ear and out the other, but at least I discovered I was capable of learning about them, in case the need should ever arise.

Why did I feel so competent in other subjects when I was a child and so stupid in math? What causes this type of "math block" anyway? Why are so many of us exasperated when trying to help our children learn simple arithmetic concepts? Does it have to be the most difficult part of teaching our children at home? Are we all doomed to failure? Must we turn it over to our husbands in frustration? Is it just a "guy" thing?

As I began the struggle all over again with my own children, I started asking myself some of these deeper questions. Gradually, the answers started coming, but it took several years of failures and further frustrations before I finally figured out what was wrong. It wasn't my fault. My kids weren't stupid or lazy. I was just using curriculum materials for math that were hopelessly flawed, because they were constructed on faulty premises concerning how children learn. The authors and publishers of these materials had ignored or misunderstood important principles concerning readiness, intellectual development, learning styles, and the effects of bad attitudes, labels, and self-fulfilling prophecies. They had simply charged ahead with their teacher-oriented lesson plans, forgetting that children don't think or process information the same way adults do.

You don't need to make the same mistakes I did. By reading this booklet before any damage is done, you will build an understanding of how your children learn and will be able to use whatever curriculum materials you select more intelligently than I did. If you feel the slightest bit creative, it may even be possible for you to design your own elementary math curriculum, using cheap materials and natural experiences instead of expensive kits and poorly-designed texts and workbooks.

As for me, I actually enjoy math these days, but I still have to have a pencil and paper or a calculator in front of me, and I still make stupid mistakes when I'm stressed out or tired. I still feel a brick wall starting to build when I hear that question about the car that starts out from Baltimore. The men that visit my booth at the curriculum fairs and add up their totals in their heads can reduce me to a bowl of Jello in two seconds, especially when they start enthusiastically explaining their systems to me.

I just can't do math like that! I'm a "pencil and paper" person, and I'll never be an "in my head" math whiz, no matter how many tricks somebody else tries to teach me. I believe that understanding and accepting such individual learning styles is an important step that we all have to take on the road to "unblocking" those math cells and developing a feeling of competence in ourselves and our children. So don't ever listen if someone tells you that math is difficult and frustrating, and suggests that you or your children are hopeless! It really doesn't have to be that way at all! In the following sections we're going to learn how to take that frustration out of math, and learn to actually enjoy it (or to at least muddle through together with a minimum of tears.)

A Few Precautions Before We Begin....

I hope that I can teach my children to become good at math and learn to enjoy it someday, but whether I do or not, I am determined that I will never make them feel as stupid or lazy as I did. I'd rather do nothing than make them hate the subject with a passion. I'll never keep on plowing if it's obvious that I'm beating a dead horse. That's the good part about being a home-schooling mother. I know that I never have to hurt my children if I just use a little common sense and recognize when it's time to back off and wait awhile.

You may not know much about math either. You may think you'll never be capable of teaching algebra, but you know one thing better than anyone else on earth. You know your own kids. You know whether something is working or not. You know when something is sinking in, and you know when the kids are floundering and feeling like morons. Trust your own instincts, and stop before any serious damage is done. You can always try again later, using a slightly different approach.

Remember, too, that life and education don't stop when the kids reach age eighteen. For some people, like me, the motivation for learning certain subjects may not come until they really need to know the information on their own. I didn't learn a speck of math until I was over twenty-one, and somehow I managed to go to college, get married, have kids, run a business, get three graduate degrees, write a couple of books, and eventually learn enough math to teach algebra to high school students. Obviously, my own late start didn't seriously hamper my ability to achieve some worthwhile goals.

You have many responsibilities towards your children, including the responsibility for understanding their developmental levels and presenting them with appropriate materials and experiences. Ultimately, however, their motivation and learning are their own responsibility, not yours. As the old adage states, "You can lead a horse to water, but you can't make him drink." You can, however, notice when your horse is thirsty, pick out the tastiest, most suitable watering hole, remove his bit and blinders, and refrain from yelling and beating him while he's bending over. If there had been a little less yelling going on when I was a child, and if someone had waited until I was really "thirsty" (ready and willing to learn), maybe I would have been better at math, too!

Children Must Build Their Own Understanding of the World

The first thing you have to understand in order to work with young children is that they do not perceive things or process information the same way adults do. Before the age of twelve, children have a great deal of difficulty thinking in abstract terms. That's one of the reasons that so many young children are incorrectly diagnosed as "dyslexic." Most children don't see letters of the alphabet as abstract symbols, like we do. Instead, they view them as pictures of real objects. Real objects can be turned around. An apple is still an apple when it is upside down, so why does an "m" become a "w" when it is upside down? An umbrella is still an umbrella when you turn it from left to right, so why does a "J" change into something else when it is turned the other way? It just doesn't make sense to them. This is particularly true with artistic children, or those with visual learning styles, who tend to focus even more on the way these symbols look, rather than on the meaning behind them.

Mathematical symbols are even harder for most young children to deal with. Kindergarten students may happily count from one to a hundred, but it is very difficult for them to connect the words and symbols with the quantities that they represent. In order to get to the point where children can understand these symbols in an abstract way, they must first handle the real items that the symbols represent. Children need to see things, to touch them, and to play with them. They need to turn them over in their hands, and ask themselves questions. That's why kids are natural scientists. They are always fooling around with objects, looking, contemplating, experimenting, and asking "Why?", "How?", "What makes this tick?"

Children can be natural mathematicians, too, if they are allowed to be. They enjoy playing with numbers, counting and sorting real objects, and discovering arithmetic truths on their own, without excessive adult interference. Nothing you "teach" them will ever click until they discover it for themselves and make it their own. I noticed this again just the other day. My daughter Laura, age 9, was in the car with me, and we passed a sign that said, "Milk, 215 cents". She was silent as we drove the rest of the way home, and then, just as we were getting out of the car, she exclaimed, "You know what? $2.15 is the same as 215 pennies!" Then she went on to figure out how many pennies were equal to $112, the amount of money she has saved for a horse. She was extremely excited, since it sounded like a lot more money when expressed in terms of pennies. (She probably thought, in some child-like fashion, that she had just gotten closer to her goal.)

I immediately wanted to scream, "NO KIDDING! WHAT DO YOU THINK WE'VE BEEN STUDYING FOR THE LAST TWO YEARS!" But I restrained myself. I knew it was just another example of the fact that children have got to re-discover the way math works for themselves. It didn't matter that I had gone over that very point with Laura a hundred times. Children have got to construct their own understanding of the world around them, one step at a time. They learn through a process that is a lot like laying bricks. One layer has to be laid down at a time. Without knowing the layer of understanding that they currently hold (which can be very difficult to pinpoint), we have no clue as to what they might be capable of learning next. That's why the main responsibility for learning lies with the child, rather than the parent. All we can do is provide the experiences, the materials, the language, and the feedback. They need to develop their own understanding of the world around them. If we think we can shortcut this process by simply "teaching" them concepts that have been developed by others, we are in for a rude awakening. Learning just doesn't work that way.

The belief that children must construct their own knowledge a little at a time, through interaction with their social and physical environment, always building on what they already understand, is known in the scientific world as "constructivist theory." The most famous of these theorists was Jean Piaget, whose ideas have had a tremendous impact on the math curricula that are currently available in both the public schools and the home-schooling movement.

Unfortunately, although these curriculum materials are supposedly built on his work, most of them are actually counterproductive, because his ideas have been mis-applied. When Piaget explained the phases that children pass through on the road to mental maturity, American educators reacted in a rather typical manner. They tried to speed up the learning process by moving all the phases down a notch, teaching concepts before the children were ready. Our culture is famous for its tendency to rush around, hurrying everybody up and pushing, pushing, pushing. Piaget laughed and referred to this phenomenon as the "American problem," but it's really no laughing matter when children are hurt in the process. Rushing their development not only doesn't work, but it causes children to fail, and sets the stage for confrontations between frustrated teachers and their equally frustrated students.

In the chapters that follow, I will explain the various phases of cognitive development that Piaget discovered, and discuss what does and does not work during each phase. However, education is always an individual affair. Although children pass through the same stages on their road to maturity, they will not all pass through them at the same time, or with the same level of difficulty. Some kindergarten students will have sufficient interest and mathematical talent to do work that others won't be ready for at the age of twelve. As usual, you need to take the information given and apply it to your own children and your own situation, using your intuition and common sense as a guide.

Who Was Jean Piaget Anyway, and Why Should We Listen to Him?

Piaget was a Swiss psychologist, who lived from 1896 to 1980. He began life as a child prodigy, publishing a major work on the albino sparrow at the age of ten and earning a doctoral degree in zoology by the time he was twenty-two. After getting married and having a family, he became fascinated with observing his children and watching how their intellectual abilities developed. He decided to switch from studying animals to studying the mental, or "cognitive" growth processes of children.

As a natural scientist who believed in the theory of evolution, Piaget grounded some of his work in the ideas of Darwin. He believed that all "organisms" (including children) start off with a certain genetic make-up, and then gradually mature and adapt to their environment through a process known as "assimilation and accommodation." In other words, babies have a certain amount of built-in intelligence at birth, and gradually develop this intelligence through a combination of growing (maturation) and interacting with the environment (learning through playing, fiddling around with things, and talking to others about their discoveries). When they stumble on a new discovery, or a new piece of information, they must first "assimilate it" (take it in), and then "accommodate" the rest of their knowledge to make the new piece fit properly.

For example, suppose a toddler has already figured out that objects sink when placed in the bathtub. Then one day, he notices that the soap that mom throws into the tub is staying on top. He must assimilate the new information. ("Hmmm - this thing isn't going down to the bottom of the tub.") He then asks the inevitable "why" question. His mother will typically say something like, "The soap is floating." This provides the language that the child needs in order to make some sense of the new experience. He "accommodates" the new piece of information by adapting his previous knowledge. Before, he thought that everything went to the bottom of the water in the tub. Now he realizes that some things sink and some things float.

As Christians, of course, most of us would disagree with Piaget concerning the evolutionary backdrop for his theory. However, we don't have to believe our ancestors were monkeys in order to recognize that many of Piaget's ideas about mental development are sound. Through scientific observation, he learned a great deal about how children think. Studying his work can help us understand how children learn about the world around them, and why they have trouble with certain concepts. That information can be invaluable in planning instruction in math and science.

Piaget's Stages of Cognitive Development

Piaget divided children's cognitive development into four stages: "sensori-motor," from birth to two years; "pre-operational," from 2 to 7 years; "concrete operational," from 7 to 11 years; and "formal operational," from 12 to 15 years. During the sensori-motor stage, the baby uses his senses and his developing motor skills to interact with the world around him. During the pre-operational stage, children primarily focus on language development and often use drawing as one of their main means for reaching out to communicate with the world around them. As you will learn in this booklet, a formal study of math is not a very appropriate task for the pre-operational child. When children move into the concrete operational stage, about the age of seven or eight, they become ready to grapple with mathematical concepts, but need to see and manipulate concrete objects in order to understand what they are doing. It is not until the formal stage of operational thought, which starts about the sixth or seventh grade, that children are really ready for abstract thought and the manipulation of symbols.

To summarize, the phases of development that Piaget outlined are as follows:

Birth to 2 years:	Sensori-motor phase
2 to 7:	Pre-operational phase
7-11:	Concrete operational phase
12 and up:	Formal operational phase

In the following pages, I will discuss each of these phases and suggest proper developmental experiences at each stage. Always remember to allow for individual differences. Rely on your own intuition and knowledge of your children. Never do anything or fail to do anything you know is important simply because some "expert" said so, even (or maybe especially) if that "expert" is me!

Sensory-Motor Development
Approximate Age: Birth through Two Years

From birth to two years of age, babies and toddlers are in the sensori-motor stage of mental (or cognitive) development. The stage is so named because babies focus on learning about their bodies and their immediate environment mainly through their senses and through physical movement. During this stage, the learning processes are advancing at tremendous speed. Babies are taking in everything in their environment, and actually creating some of the biological structures in the brain that will later determine their level of intelligence and capacity to process information.

Because of the importance of this early period, there has been a great deal of confusion about the proper way to treat babies. Some advocate pushing both physical and intellectual development and creating "brighter babies." We've all seen the "gymborees" and "brighter baby institutes" advertised, and wondered if they really accomplish anything meaningful. Glenn Doman, one of the advocates of this type of early learning, has written a book entitled *Teach Your Baby Math*, which encourages mothers to make flash cards with dots in order to help their babies develop the ability to recognize complex numbers and operations early in life.

There is no question that babies are capable of tremendous development during the first two years of life. The question that divides knowledgeable people is the question of methods. What is the appropriate way to stimulate the highest possible level of development? Is there anything that people do that can actually harm or thwart that development?

Personally, I believe in a natural developmental process. Flash cards and intense learning experiences may "work" in some cases, but they aren't usually necessary, and may actually be counterproductive in other areas, such as emotional development. However, it is certainly possible to thwart physical and mental development during this stage. Children who spend most of their time in swings and plastic seats don't have the opportunity they need to exercise their growing sensori-motor skills. Those who are left alone in cribs or playpens do not have sufficient stimulation or contact with people to develop their mental processes or to start learning language and interaction skills. It's possible that children growing up in these situations might need the extra stimulation of identifying flash cards for an hour of "quality time" a day when their mother picks them up from day care. However, when a mother takes her child around with her everywhere she goes, like most homeschoolers do, the child is exposed to the sights, sounds, and experiences of human interaction on a daily basis. In most cases, this will provide all the stimulation babies need for optimal sensory-motor development. Learning how to do math can wait.

Pre-Operational Stage
Approximate Age: Two to Seven Years

The Concept of "Operations"

In order to understand Piaget's theory, we must first understand what he meant by an "operation." In Piagetian terms, an operation is something that happens when you take one or two items and do something to them that requires mental processing of some kind. For example, measuring is an operation, because you are taking an object, and using another object (such as a ruler or a scale) to figure out how long it is, or how much it weighs. Adding is an operation, because you are taking two different quantities and putting them together to get a new quantity. Dividing is an operation, because you are taking one quantity and splitting it up into several parts. Subtracting and multiplying are also obviously operations, and so is "telling time."

Consider those examples of an "operation." Now consider the name of the phase that we are currently studying, which lasts until the approximate end of second grade: **pre-operational**. What do most teachers expect children to do during kindergarten, first and second grade? They are expected to master addition, subtraction, measurement and telling time. These pre-operational children are being forced to perform operations as the main part of their math curriculum! Even worse, they are usually expected to do these operations using symbols in a workbook, without having any concrete objects around to "play with."

Teachers even try to stop them from using their fingers, which are the only concrete items they have on hand. Is it any wonder that so many children have difficulty with these concepts in the early grades?

I compare this approach to education to a parent who is trying to toilet train a young child. Mom can put an eighteen-month old on the potty every morning after breakfast, and if she is lucky, the child might start going, but the child isn't the one who is trained. The mother has trained herself to recognize the child's schedule. If she had just waited until the child was ready, at or around the age of three, she probably wouldn't have had to do anything at all.

Similarly, a person can "teach" operational behavior to young children, and wind up with children who may be able to measure objects and tell time by the age of six, but, for most of them, these operations amount to little more than "tricks." With no solid basis for their understanding, the kids are simply acting like trained animals, and will be incapable of building on that information at a later date. Remember that there are always exceptions. There will be some children who can perform some of these operations on their own at this age and actually understand them. They are usually the ones who have somehow managed to construct their understanding on their own, using real objects and activities in addition to (or in spite of) their experiences with their teachers or parents.

Characteristics of the Pre-Operational Child

According to Piaget, there are several kinds of thinking and processing that are difficult for pre-operational children. The child who has not yet reached the age of concrete operational thought has a great deal of difficulty focusing on more than one attribute of an object at a given time. One of the most important experiments that he conducted had to do with a pre-operational child (who was about the age of five) watching an adult pour water from one flask to another. One flask was tall and thin, and the other was short and fat. Both actually contained the same amount of liquid. The child would watch the adult take the taller container and pour all the liquid into the smaller container, and then watch him take the same liquid and pour it back into the taller flask. Despite the fact that the child had just seen the same amount of water in both containers, he or she would usually say that the taller one had more water in it. Why? Because the child focused completely on the height of the containers, since that was the dominant characteristic.

The ability to focus on more than one attribute at a time usually develops slowly, through normal maturational processes. Several math games, such as Cathy Duffy's "Attribute Bingo,"(part of her *Beginner Math Mouse Games*) have been designed to help students learn to focus on more than one attribute at a time. Such games may be helpful if the children are allowed to play with them on their own, without any pressure or teacher-directed lesson plans designed to push them onto the next stage prematurely. However, such games are not absolutely necessary. If allowed to mature normally, most children will pass through these stages with minimal help.

In another experiment, Piaget laid several ropes in front of a pre-operational child. The ropes were all of different lengths. One was stretched out to its full length, and the others were coiled in various positions, but they all wound up starting in the same place and ending in the same place. Most of the children ignored the fact that some ropes were coiled, and stated that they were all the same length, because they began in the same place and ended in the same place. This shows us how silly it is to try to teach measurement to a young child, especially in a workbook. If a child can't even perceive the difference between a coiled and a stretched rope, how can he be expected to measure items in a workbook? If the textbook authors were careful to always line the rulers and the items up properly, the children could probably get the right answer, but they still wouldn't understand what they were doing. As it is, the authors have taken Piaget's ideas and tried to "teach" measurement by purposely throwing in coiled ropes and measurement sticks that don't align with the items being measured. This is a clear example of an attempt to "teach" children concepts too early. These same concepts would develop naturally, with no need for teacher-directed lesson plans, if the children were just given time to mature and develop them on their own.

Practical Implications for Parents of a Pre-operational Child

How can you apply this knowledge about the pre-operational child to your math curriculum for children from the ages of two through seven? Here are several suggestions:

1. Concentrate on helping children develop a good, solid, real understanding of numbers and quantities. Counting skills can be cute and impressive, but children may be able to parrot the numbers all the way up to 100 without an underlying understanding of quantities. This is developed gradually in real-life situations: setting the table for five people, picking out a dozen apples, counting pennies that are received as a small allowance. Games can also be helpful, providing that rules are bent or dispensed with as needed. Texts and workbooks are totally unnecessary and frequently counterproductive at this age. If you decide to use them, choose them carefully, using the guidelines found in the next section.

2. Consider giving children personal clocks or watches to encourage them to learn to tell time gradually. I recommend starting with a digital watch, which is easier for most preschoolers to understand. Wait until about second or third grade to introduce an "old-fashioned" clock face.

3. Begin slowly building a math-related vocabulary. Use words *in normal conversation* like greater than, less than, more than, equal to, etc. Later start using words like "plus," "minus," "sum," and "remainder," when using math in real life or when playing games that introduce concepts such as addition and subtraction.

4. Use real-life experiences to introduce measurement concepts. When gardening, measure the distance between plants, rows, etc. Decide together how many plants you are going to put out, and then count them out together. When cooking, introduce various measuring devices such as measuring spoons, liters, cups, etc. (Use one-cup measures, if possible, instead of two-cup measures which tend to be confusing.) When shopping, try to involve the children in a natural way. Weigh produce together, count out 16 oranges, ask for help with coupons, etc. There are many other real-life uses for math: laying carpet, putting up wallpaper, carpentry work, doing taxes, bookkeeping, balancing checkbooks, handing out allowances, saving or tithing. As much as possible, involve your children in these real-life experiences. However, always consider their interests, attention spans, and readiness levels, and don't push experiences on them if they are clearly resistant. Usually children aren't being disobedient or lazy. Your timing is probably just off, and their heads feel like they are stuffed with marshmallows.

5. Consider setting up a "math center" in your home, with a number of math games and manipulatives that are down low and accessible to the children. Establish a few ground rules, and then give the children freedom to use the materials when the spirit moves them. As usual, you could spend a lot of money stocking this center, but it is also possible to start with cheap and free items, such as beans, popsicle sticks, Cheerios, and home-made games. The more expensive items can be added as the pocketbook allows.

You might want to include various items for counting, sorting, or sequencing. Old egg cartons work well for these experiences. Sort by categories (color, size, type.) Sequence from left to right; heaviest to lightest; first to last; biggest to smallest. Also include some math-related games, like *Memory* (which builds concentration skills), Dominoes, Cathy Duffy's *Beginner Math Mouse Games*, and *Number Neighbors*.

However, remember that children have a hard time using rules, so let them use the materials in any way that makes sense to them. Sometimes the mistakes they make teach them more than the game would teach them if played "correctly." (For example, my five year old tried to play *The Neighborhood Game*, in Cathy's set. It is supposed to teach addition, but he tried to match the numbers on the board with those on a single die, and quickly discovered that he didn't have all the numbers he needed. When he came to me, thoroughly frustrated, he was READY to learn a little bit about addition, and the use of two dice.)

6. Consider getting an abacus for young children to play with and placing it in the math center. I personally believe that children get much more benefit out of this tool at an early age than they do later on. They will learn concepts like "one more than," "odd and even," and beginning adding, subtracting, and multiplying skills.

7. As children turn "school age," remember that there is no "magic line" that is passed over that indicates that instruction needs to mutate into something with more structure. Everything is simply a continuation of what they have already been learning.

8. When children are six or seven years of age, help them develop an understanding of "sets." These are groups of things that have something in common: a set of five apples; a set of all the black things in the world. Define a set together and then decide what things belong in it and what things don't belong in it. Try to help them understand that mathematical sets should have clearly-defined borders. It's harder to agree on what constitutes a set of "boring subjects," than a set of "winged creatures," isn't it?

9. Continue helping them develop a real understanding of quantities. Work with numbers up to 100 first, then gradually help them understand and work with numbers up to 1000. We used to play games in the car, asking the children to count by two's, or five's, or three's. Then we would have them count backwards by these quantities. (This is much more difficult!) Such exercises are good practice for multiplication later on.

10. Place value is a very abstract concept and it may take a long time for children to really understand this idea. Consider getting or making place value cards to help children understand. These are long, overlapping strips with numbers on them. Directions for making cardboard cards can be found in the book, *Activities for the Abacus*, by Joan Cotter.

11. Most children will begin to learn about addition and subtraction on their own, while they are playing math games or working at real life activities (like playing store or baking or counting their own money). Don't start worrying about the use of flash cards or "learning wrap-ups" in order to "get faster" just yet. That can wait until third or fourth grade.

12. When the children seem to have a good understanding of adding and subtracting real numbers, and can also understand a little about place value, you may want to begin helping them learn about adding and subtracting with more than one digit. Remember to use real-life activities as much as possible, and add math games or concrete activities a little at a time. (Don't expect a second grader to be able to add or subtract two and three digit columns in a workbook unless he or she has already developed this ability using concrete objects.)

One activity for developing this knowledge is to make a large poster that is divided into three (or more) columns. Label them from right to left, *Ones, Tens, Hundreds*, and use them for teaching "carrying and borrowing." Lay the paper down on the table, and use bundled popsicle sticks to teach these concepts. For example, if you were going to tackle the problem "25 plus 37," you would put two bundles of ten sticks in the tens column and five loose sticks in the ones column. Then you would put seven more loose sticks in the ones column, and three more bundles of ten in the tens column. Finally, you would start adding, beginning in the ones column. When the children notice that there are more than ten sticks in the ones column, suggest that they bundle up ten and put them in the tens column, and help build their math vocabulary by introducing the term "carrying." Use the same idea to teach "borrowing" in subtraction. I always taught these skills with a smaller sheet of lined paper on the side, and had them do the problems on the big sheet first with the popsicle sticks and then write them out afterwards.

Using Textbooks and Workbooks with the Pre-operational Child

As you can probably tell, I'm not a great fan of using textbooks and workbooks at this age. However, if you have already bought some materials, or your children really enjoy using workbooks, you may want to incorporate them into your curriculum. In the next section, I list some guidelines for choosing and using textbooks and workbooks with pre-operational children.

Guidelines for Selecting Primary Texts & Workbooks

1. Choose books that are colorful and look like they are fun. Don't assume that workbooks put out by curriculum publishers are automatically superior to those found in discount stores. This is not true.

2. Preschoolers and primary students usually prefer a workbook that they can write in to a textbook that requires copying the problems onto a separate page.

3. Try to select a workbook that has one type of problem per page. Textbook authors are always acting like it is some major accomplishment for children to "take notice" of the minus sign or the addition sign, but mixing up the problems usually winds up confusing them instead.

4. Use a workbook that has helpful pictures. If you can't understand what the picture is trying to show, how will your child understand?

5. Consider taking individual pages out of the book and placing them in the math center. Large books can sometimes either overwhelm a child or make those that are perfectionists feel like they have to finish every page in a given year or be failures.

6. Watch out for problems (and sometimes entire chapters) that are not appropriate for this age group, and skip over them. (Even if the textbook is supposedly geared for your child's grade level, remember the publishers probably are trying to push Piaget's stages down lower than they should be.) Before the age of eight, avoid the following:

Turned around problems. (5 + __ = 8) Many young children see the five and the eight and the plus sign, and wind up with an answer of thirteen. Pre-algebra experiences can wait until the kids are pre-teens.

Number lines.

Problems involving measurement or time. (Set up your own real-life learning experiences instead.)

Timed exercises, which place unnecessary stress on young children.

Tests and grades. You know what they understand. Why make them prove it?

Using Manipulative Programs

Some curriculum publishers are starting to understand the need for young children to play around with concepts, rather than emphasizing the use of pencil and paper problems. A number of manipulative programs have been designed and marketed in recent years. In general, these are usually designed better than textbook series for young children. However, the mere fact that they are manipulative-based does not always make them appropriate. You still need to use good judgement. In most cases, children who are in the pre-operational stage will do better learning on their own, with a little help from you, than they will learning through teacher-directed experiences. A major drawback of some of these programs is the expense involved. You can often create your own cheap manipulatives that are just as effective as the expensive variety. There is nothing "mystical" about colored rods or tiles. Children can learn just as well with any tangible, concrete objects that can be counted, sorted, grouped, and manipulated.

Concrete Operational Stage
Approximate Age: Seven to Eleven Years

By now, you are probably familiar enough with these terms to figure out what this stage involves. Around the age of seven or eight, something seems to click inside children. That's probably one of the reasons that Raymond Moore, author of *Better Late than Early*, believes that formal instruction should be put off until this time.

When children move into the concrete operational stage of cognitive development, they become capable of performing fairly complex operations. These include most of the operations that the textbooks for third through sixth grade include: addition and subtraction with regrouping (carrying and borrowing), multiplication, division, work with fractions and decimals, advanced measurement, other number systems, and so forth.

There is still a problem with most of the available textbooks, however. Notice there is another word in the name: **CONCRETE** operational. The children at this phase of mental development still need to do things hands-on before they can really understand what they are doing with symbols on paper. Unfortunately, American educators typically drop all the concrete experiences at the exact age when they are needed most.

Actually, this is the age when formal instruction in math should **begin**. I included some ideas about working with your children in the early grades because most of you probably won't feel comfortable leaving them alone until this age. However, if you could see your way clear to postponing formal math instruction entirely until your kids reached the concrete operational stage, your job would be a lot easier! (Of course, younger children will learn a lot on their own. They will be encouraged and helped by the provision of materials and experiences and by your natural interaction with them at a younger age. I'm just talking about postponing **formal instruction** with them until the age of seven or eight, not about forgetting to play with them and involve them in the natural processes of learning through living.)

Characteristics of the Concrete Operational Child

As I have already stated, the child who has moved into the stage of concrete operational thought is now ready for formal instruction in math. However, he still needs concrete experiences in order to make sense of the new material. A total reliance on textbook instruction is going to lead to frustration.

Between the ages of seven and eleven, most children are very good at following rules. In fact, most of them are "hung up" on rules. That is why it is so difficult for children this age to play games with younger children. The fourth grader *lives* by the rule book. The kindergartner or first grader either doesn't understand the rules, or wants to be creative with following the rules (i.e. changing them to suit his or her own purposes). The end result is usually two children who are fighting and arguing the entire time they are together.

Because children this age are so good at following rules, they can often get right answers when using textbooks, even though they don't have the foggiest notion what they are really doing. A typical fourth or fifth grader can flip fractions when dividing them because "that's how it is done," and can do long division perfectly and quickly (bring down, divide, multiply, subtract, bring down, etc. etc. etc.). When asked what they are doing, very few of them can explain anything clearly. There's a good reason for that. They don't have the slightest idea!

When children are allowed to breeze through textbooks, covering material and getting right answers without understanding what they are doing, they are not laying those "bricks" of understanding that they will need for later work in algebra and geometry. It is absolutely crucial that they still be given the opportunity of "mucking" around with math so that they can really learn the material before they are allowed to move on to more difficult material.

Practical Implications for Parents of a Concrete Operational Child

This is the age when a good series that combines manipulatives with written work might be of value. If the cost of such a manipulative program is prohibitive, however, it is still possible to use a textbook series and add the concrete experiences that are needed.

1. Begin by sitting down with your students and trying to assess exactly where they are at the present time. Don't be satisfied with having them *show* you what they can do on paper. Ask them to explain it. Give them something to manipulate, like cheerios or raisins, and ask them to demonstrate what they are doing when they are multiplying, performing long division, or working with fractions.

2. You will probably find that most students can explain clearly how to add, subtract, and multiply. A few will understand division, although very few will be able to explain the long division process. (Frankly, I'm not sure I could do a good job, either.) If they can clearly understand the basic concept of division, it may not be crucial that they can explain long division, as long as they can perform it.

3. Many students will be unable to explain or perform concrete operations using fractions and/or decimals. If you determine that your students don't really understand these concepts, it is imperative that

you back up and create experiences where they can play with fractions and decimals until they begin to understand the operations they are performing. There are many commercial games available that help teach fractions. Fractions can also be learned in the kitchen or the workshop. Decimals can be best understood by fooling around with money and time. (Why is a quarter hour fifteen minutes and a quarter 25 cents?) Decimals can also be used in the workshop.

4. One of the reasons that many students this age hate story problems is that story problems often require a genuine understanding of the concepts. The students can follow their "rules" easily to get the right answer when the problem is already set up for them. However, when they have to set the problem up themselves, their lack of understanding shows. Many of the textbooks available to home schoolers lack sufficient practice with story problems. One of the best ways to expand the thinking ability of your children is to ask them to make up story problems themselves. You might begin by asking older students to make up story problems for younger students. Eventually, they will begin to understand the idea behind story problems, and work up to creating their own story problems for the level they are working on themselves. At that point, true understanding will develop. One good resource for working on this skill is *Math by Kids*, a compilation of story problems written by children that is edited by Susan Richman. Working on story problems that have been written by other children may encourage your children to create similar problems themselves.

5. This is the perfect age to begin working on geometry. Because they are in the concrete stage of operations, children will enjoy working with their hands to create various geometric shapes. Ask them to "build" a square, a cylinder, a cone, a rectangle, etc., using cardboard or construction paper. Don't give them any instructions. Just show them what the shape is supposed to look like, and leave them alone to figure it out. *Hands-On Geometry* by Activities for Learning is a good tool for working with children of this age. The booklet helps children create a variety of geometric constructions (drawing angles, perpendicular and parallel lines, etc.). The kit itself could also be gathered from materials available at most discount stores. Children should learn to use such materials as protractors, compasses, English and metric rulers, and maybe even a slide rule. This is also a good age to play with "tangrams." Tangram kits can be purchased at many teacher stores, and there are instructions for making tangrams in the *Family Math* book which is sold by several home school suppliers.

6. This is also a good age for playing math games. One of the best I know is *Muggins*, which is a fast-paced game involving multiplication and division skills. One of the reasons I like it is that fathers often enjoy playing with the kids! A variety of home-made games are also featured in the *Family Math* book. This book was written by math professors at the University of California, and many "higher level" thinking skills are called into play.

7. Children in the upper elementary grades should also be encouraged to use math in their daily lives as much as possible. If there is a family business, begin to get them involved. My children started working in my curriculum booth at this age, and it really helped their math skills to learn to add up orders, figure tax, etc. If you don't have your own business, perhaps they could start a small business of their own! My older son once had a "pet sitting" business for awhile, and my daughter has earned money babysitting and working for me. (Among other important things, it has taught her she doesn't want to be a secretary!)

8. Other "real-life" experiences should continue at this age. Gardening, woodworking, baking, etc. can now help enormously to add "life" to the problems they are working on in their textbooks.

Using Textbooks or Workbooks with the Concrete Operational Child

By this age, most children can use textbooks or workbooks, provided that hands-on experiences are either built into the series, or added by parents. A few precautions are still in order:

1. Be responsive to your child's needs. Textbook series have different strengths and weaknesses. Saxon math worked great for my oldest son, but my oldest daughter totally bombed when trying to use the same series. Saxon usually works well for students who are intuitively good at math. For students who have more trouble, it may not add in enough "fooling around" with concepts. If one series isn't working, search until you find something that will.

2. Don't get hung up on "grade level." Determine what your child really understands, and start from there. If the child needs a third grade book in the fifth grade, so be it! Don't feel like either you or your child is a failure if you don't finish a particular book in a single year, either. We often took two years to finish a single textbook.

3. Don't force your children to do endless repetitions of the same material. If a student can do two long division problems, he doesn't need to do fifteen more for practice.

4. Although endless repetitions aren't necessary on a daily basis, it is helpful to build in constant review of material that has been covered earlier. This is one of the biggest strengths of the Saxon series. Other series may be weak at providing adequate review. If this is the case, it is up to you to supplement with sufficient review so your children don't forget material that they have already mastered.

5. If children really understand material and want to move on, let them skip pages. At the beginning of each Saxon book, there is a tremendous amount of review because most school children have taken off the entire summer and forgotten a lot of what they knew. Since my children didn't take that much time off, they hadn't forgotten as much. I let them do every fifth lesson until they got to a part that started getting difficult. That way, they weren't bored to death for ten weeks waiting to get to something new.

6. Some children enjoy using workbooks and dislike texts because the work of transferring the problems to the page tires them out before they even begin. If this is the case, consider a few alternatives. They can either continue to use workbooks (one series for older children is the Key Curriculum Press series) or you can write out the problems for them on a sheet of paper. You might also consider using a computerized math program for children who enjoy computer learning more than textbooks.

7. If you don't feel competent in math, and your children can't seem to learn using textbooks, consider getting a video series to help. Some children learn better visually, and need to actually see an adult explain things. However, be careful when using video programs, because they can have the same effect that textbooks often have when used in the absence of real-life experiences or manipulatives. Kids can "learn the rules" without developing any real understanding of the processes involved. Children may sail through the videos with no difficulty, but lack true comprehension of the underlying material.

Formal Operational Stage
Approximate Age: Twelve Years and Older

Around the age of twelve or thirteen, children develop mental maturity. Naturally, this doesn't mean that they grow up overnight. They will be gathering new knowledge and growing in wisdom and understanding for years to come. However, the basic mental structures are now in place for formal, abstract thought to begin. That's why teenagers are so anxious to discuss complex societal issues, and why they so often come up with ideas that don't fit in with the views of the rest of the family. They are merely exercising their new-found ability to reason and think about abstract issues on their own. Don't worry too much if they seem to go off the deep end on certain issues. Sanity will return with time.

At this point, if your teenagers have been allowed to construct their own understanding during the elementary years, and have had sufficient real-life experiences and concrete materials to work with, they will be ready for the formal study of mathematics using abstract symbols and textbook approaches. *However*, if they have not laid down the proper building blocks, they will not be ready to proceed until these deficiencies have been corrected.

In other words, don't start with algebra until you are convinced your students thoroughly understand basic math! Remember that understanding is not the same as being able to get the right answers. If their understanding still seems to be shaky, put away the algebra textbooks until they get the basics down pat.

Many parents wonder whether all students need to take algebra and advanced math courses. The answer depends on the student's future plans. For those planning on going to college, the SAT and ACT require a high level of advanced mathematical knowledge. For those whose plans don't include college, an emphasis on consumer and business math may be more appropriate. If their plans change later, they will probably be old enough and mature enough to make up any deficiencies on their own. That was what happened with me.

Remember, too, that students have different needs when it comes to learning, even at this age. As I mentioned earlier, my oldest son sailed through the Saxon math series with no problem. My daughter, who tends to be a visual learner, floundered until we ordered the Chalk Dust video series. In one year, using this series, she raised her score on the SAT from a 47th percentile to a level that won her an academic honors scholarship. In the next section, we will be discussing the need for recognizing and responding to such differences in learning styles when working with individual students.

Other Considerations When Teaching Math

Learning Styles

Most of you are probably already aware that there are at least three distinct learning styles. Some students are ***auditory learners***, who tend to take in information with their ears. These are the students who rarely need to take notes, and who prefer to listen to tapes and speakers instead of reading material for themselves. They are also the ones who learn to read well using phonics. ***Visual learners*** need to see information, to take notes, to use outlines, and to see the speaker who is explaining the information. That's why Ginny, my "visual" daughter, is doing better with a video series in algebra. Teachers use chalkboards and visual props, such as overhead projectors, in order to adequately reach such visual learners. Other students tend to be ***kinesthetic learners***, who need to manipulate objects physically in order to understand. Most children have a basic need to manipulate concrete objects during the earlier years, but kinesthetic learners are even more oriented in this direction. Even as adults, kinesthetic learners tend to perform best in areas such as mechanics, plumbing, carpentry, and crafts.

In math, there are also recognizable learning styles, which follow a similar pattern. I mentioned earlier that I tend to be a "pencil and paper" person in the area of math. This usually corresponds to a visual learning style, which seems odd, because I'm really an auditory learner in all other areas. It's not that unusual to find people with such "mixed" styles, however. Some of us tend to learn one way in math and another way in areas of the curriculum that are dependent on language skills.

People who tend to be auditory learners often are good at doing math in their heads. My oldest son falls into this category. (Needless to say, I miss him at my curriculum booth now that he is busy with his own pursuits.) One of our biggest battles during his younger days was over his desire to "just put down the answers." I eventually learned to let him do things in his own fashion, as long as he could explain his thinking process clearly. (Not that I usually understood what he was talking about, but I could tell that he understood it, and that was the important thing.)

The kinesthetic learner in math is even more dependent on manipulatives than other learners. When kinesthetic children are learning math, they absolutely have to have concrete props to count, sort, divide, and multiply. Before they can do problems on paper, they must physically manipulate objects until they can understand the concepts involved. In extreme cases, they may be unable to write down the problems on paper until they reach the teen age years. If they can demonstrate an understanding of the concepts, there is probably no reason to try to force them to use pencil and paper.

Male/Female Differences

When I was in graduate school, one of my favorite classes involved the study of the brain. I found it fascinating to learn how the different structures and functions of the brain related to the educational needs of children. I learned that males and females actually have different amounts of brain cells designated for language and math (spatial) skills. You can probably guess who got what when God was passing out these cells. Females have a much larger section of the brain that is set aside for the development of language and communication skills, while males have a corresponding larger section for spatial skills.

What this means, on a practical level, is that many boys will learn math easier and earlier than many girls. This certainly doesn't mean you should discourage your girls from learning math, or going into math-related fields. It just means that, on the average, boys will often have an easier time in this area, while girls tend to be better at communication skills, especially at an early age.

Learning Trends: Forward Surges, Backward Plunges, and Plateaus

When children are allowed to learn gradually, in a natural manner, with a minimum of teacher-directed instruction, they tend to learn in a somewhat uneven fashion. (I'm not saying this is bad. Over the course of time, they tend to be quite balanced. Their knowledge just develops in a rather lopsided fashion, because children tend to concentrate on one area at a time when allowed to do so.)

My middle son is a perfect example of this type of learning pattern. When he was about to go into first grade, he saw me putting a book on a high shelf and asked what it was. I explained that it was his first grade math book, and that I would work with him in it the next fall. He said, "It's mine?" and asked to see it. For the next three weeks, he ate, slept, and did math. He finished almost all of the book, and then didn't do any more math for about a year and a half!

It is very typical for students to come to a certain stage of readiness, and make remarkable progress for a period of several weeks. Then, they often reach a "plateau"—a certain level of understanding that requires that they stop and let their new knowledge slosh around inside their brains for awhile before they go any further. Sometimes, they can even look like they are taking some giant steps backwards. These are usually times when they were applying rules without understanding. At some point, they realize that they don't really comprehend what they are doing, even though they have been getting the correct answers. At such a time, they may feel the need to backtrack for awhile in order to build the knowledge that was lacking.

These forward surges, backward plunges, and plateaus can be difficult to tolerate if you believe that all knowledge develops in a straightforward manner, and if you expect students to show progress in every area of the curriculum every year. One year, my son did nothing but social studies. During that year, he learned more about geography, history, and the biographies of famous people than I have ever known myself. During that year, I also had to bite my tongue to stop myself from hounding him about his math, writing, and the other areas of the curriculum. This year, after taking an entire year's hiatus from math, he is once again making huge forward strides, and is almost ready, (at the age of twelve) to move on to algebra.

You Are the Expert

I realize that I've presented a lot of material in this booklet, and you may be wondering how you should proceed at this point. Really, it isn't as difficult as it may sound. Always remember that you have an advantage over any other so-called expert in any field. You know your own children better than anybody else. You know in your heart when something is working and when it isn't. If everything is going fine, great! (But do be sure to check that the children are really understanding what they are doing, and not just following rules and getting correct answers.)

If something hasn't been working well, back off for a while and try to figure out why. Is it a question of readiness? Are you using curriculum materials that were poorly designed for a specific age group? Are the children taking off on some other learning tangent and simply desiring a temporary break from math? Or might it be a problem with learning styles? Would your student do better switching to a hands-on approach, or a video program, or a computerized math package?

Above all, try not to worry. Many math problems are created by frustrated teachers and parents who develop poor attitudes that are transferred to the kids. It is better to do nothing than to create problems and destroy relationships. Relax and enjoy your kids! If you can learn the same things baking and gardening together, or playing math games, why continue to plunge headlong through the workbooks amidst a flood of tears?

On the other hand, if your children love texts and workbooks, and can demonstrate a true understanding of what they are doing, there's nothing inherently wrong with that approach. Education is always an individual affair, so if the material in this booklet doesn't seem to apply to your family, disregard it! Never, never, never stop doing something that is working well for you simply because you read somewhere that your approach was wrong. You are the only expert that really counts, after all, and you're probably already doing a wonderful job with your children. Isn't that the most relaxing thought of all?

Resource List

Alpha Plus
P. O. Box 185
Chewsville, MD 21721
(301) 733-1456
(general manipulatives, etc.)

Chalk Dust Company
11 Sterling court
Sugar Land, TX 77479
1-800-588-7564
(algebra I & II, geometry,
SAT review video courses)

Cornerstone Curriculum Project
(David Quine's material)
2006 Flat Creek Place
Richardson, TX 75080
(214) 235-5149

Cuisenaire Company
P. O. Box 5026
White Plains, NY 10602-5026
1-800-237-3142
(manipulatives)

The Elijah Company
1053 Eldridge Loop
Crossville, TN 38558
1-888-ELIJAH
(carries many of the math programs mentioned, including Saxon math, *Math Mouse* games, *Muggins Math*, *Family Math*, plus lots of other teaching materials for home schoolers and concerned parents)

Key Curriculum Press
P. O. Box 2304
Berkeley, CA 94702
1-800-338-7638
(workbooks for all ages)

Learning Wrap-Ups
2122 E. 6550 So.
Ogden, UT 84405
1-800-992-4966

Math Teacher's Press
(Moving With Math-manipulatives series)
5100 Gamble Drive
Minneapolis, MN 55416
1-800-852-2435

Math-U-See
(another manipulative series)
1378 River Road
Drumore, PA 17518
1-800-383-9585

The Relaxed Home Schooler (Me!)
P. O. Box 2524
Cartersville, GA 30120
(general games, books & resources)

Saxon Publishers
1320 West Lindsey
Norman, OK 73069
(800) 284-7019